Snakes of Virginia

A Kids' Nature Guide

Discover the Scaly, Crawling Creatures Around You

Eastern Copperhead

Josiah Fogle

Let's learn
about snakes!

Eastern Wormsnake
© John White

Snakes of Virginia

A Kids' Nature Guide

Discover the Scaly, Crawling Creatures Around You

Josiah Fogle

FOGLEWORKS

To my children,
Liam, Caleb, Tyler, and Madelynn.

May you never lose your wonder for the world God made.

May you be brave enough to explore, wise enough to be careful,
and curious enough to keep asking questions.

This book is for you.

Contents

Contents

Hey there,
young explorer!

Have you ever seen something long and wiggly slither across a trail? Maybe it surprised you. Maybe it made you jump! Many people are afraid of snakes. After all, they don't have legs. They have shiny scales. Some even have fangs!

But here's a big **secret**... not all snakes are dangerous.

In fact, most snakes in Virginia are harmless and would rather glide away than cause any trouble. They aren't mean or scary — they're important helpers in nature!

Together, let's explore the amazing snakes of Virginia! By the end of this book, you'll learn how to tell the difference between different kinds of snakes and understand why each one plays a special role in the wild.

But before we begin, remember the most important rule of all: **BE CAREFUL**.

Now... are you ready to start your **slithery adventure**?

Common Rainbow Snake
© Terry Vandeventer

Snake Saftey!

Snakes are wild animals, so the safest rule is simple: look, learn, and leave them alone.

If you see a snake, stop and slowly step back. Never touch, poke, chase, or try to pick it up—even if you think it is harmless. Tell an adult right away and give the snake plenty of space.

When hiking or playing outside, watch where you step, especially near logs, rocks, tall grass, leaf piles, or water. Snakes like quiet hiding places.

If someone is bitten, stay calm, move away from the snake, tell an adult immediately, and get medical help. Do not try to catch the snake.

Be curious. Be brave. Be careful. Watch snakes from a safe distance!

What is a snake?

A snake is a long, legless reptile that slithers instead of walks. **Reptiles** are cold-blooded animals, which means their bodies warm up in the sun and cool down in the shade.

Snakes don't have legs, but they can move in amazing ways! They use **strong muscles** and their belly scales to push against the ground and glide forward. Some snakes can swim, some can climb trees, and some even burrow underground.

Snakes are covered in smooth, shiny **scales** that protect their bodies like tiny shields. They don't have eyelids, so it always looks like they're staring — but really, they just can't blink! And instead of chewing their food, snakes **swallow** it whole.

Most snakes are shy and would rather hide than be seen. They are important helpers in nature because they eat pests like mice and insects, keeping the environment balanced.

Snakes may seem mysterious, but once you learn about them, they can be fascinating and fun to discover!

Eastern Hog-nosed snake

Parts of a snake?

Venomous

Timber Rattlesnake

Crotalus horridus

Timber rattlesnakes are large, venomous snake. Timber rattlesnakes are usually a mix of yellowish-brown, cream, brown, and black with zigzag-shaped crossbands. These venomous snakes are the longest venomous snakes in the state. They are primarily predators of mammals but will eat some frogs and birds.

Head detail

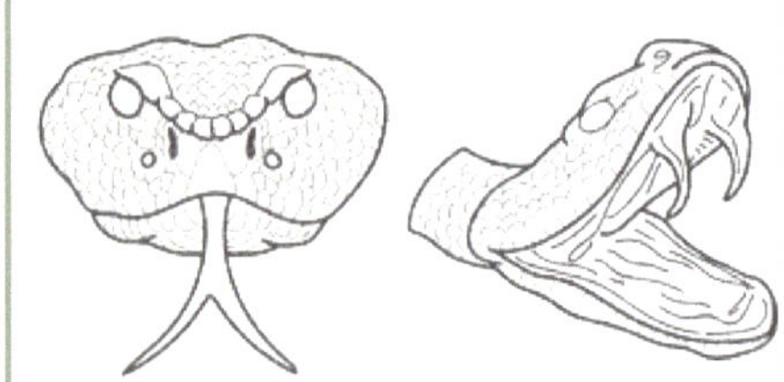

Timber rattlesnakes have a distinctive, triangular-shaped head that is wider than the neck, a key characteristics of a venoumous snake.

Habitat & Behavior

They like the rocky outcroppings and ledges typical of the mountain regions in Virginia. They prefer not to be seen, but if you do come across one move slowly and carefully. A timber rattlesnake usually won't strike unless it feels threatened.

Trivia

Range: Virginia

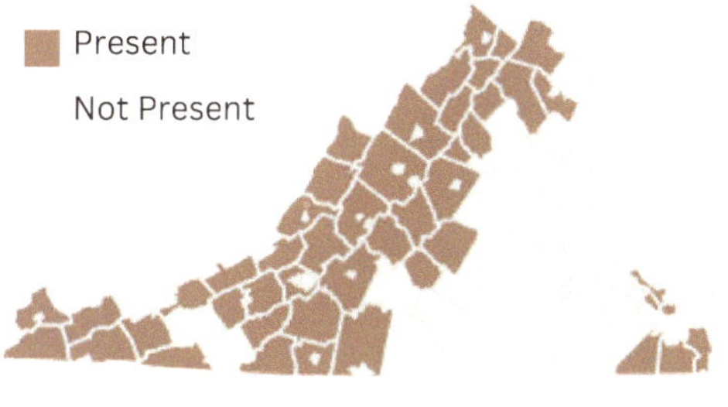

Family	Viperidae
Average Length	30 - 60 in. (90 - 152 cm)
Record Length	67.1 in. (170.5 cm)
Reproduction	Viviparous
Other common names	Rattlesnake, Banded rattlesnake, Diamond back

Venomous

Eastern Copperhead

Agkistrodon contortrix

This snake has copper-colored head and beautiful patterns shaped like hourglasses along its body. The Eastern Copperhead likes to rest quietly and stay out of trouble. Instead of chasing its food, it waits patiently for small animals like mice or insects to come close.

© John White.

Scale details

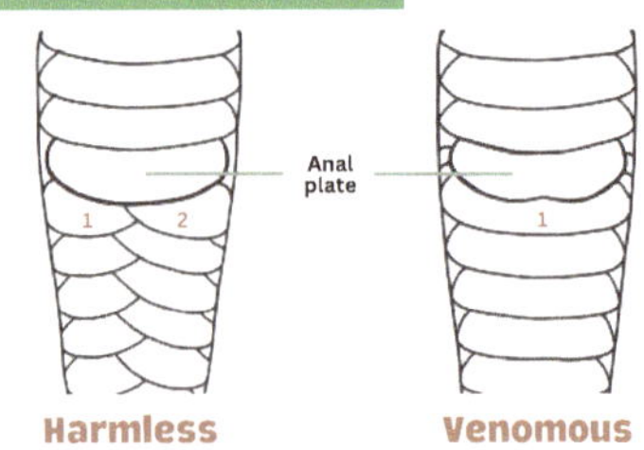

Another characteristic of all Virginia's venomous snakes is the single row of scales on the underside of the tail after the anal plate (vent).

Behavior

Eastern Copperhead is viviparous meaning it gives birth to live baby snakes instead of laying eggs. Even though it is a venomous snake, the copperhead is not aggressive.

Pop Quiz

Which snake is the Easter Copperhead?

Answer: __________

Range: Virginia

Present

Not Present

Family	Viperidae
Average Length	24 - 36 in. (61 - 90 cm)
Record Length	38 in. (134.6 cm)
Reproduction	Viviparous
Other common names	Dumb rattlesnake, Red adder, Copper adder

Venomous

Northern Cottonmouth

Agkistrodon piscivorus

Northern Cottonmouth is an excellent swimmer. It can glide through ponds, swamps, rivers, and marshes, Its body is thick and powerful, with dark bands that help it blend into muddy water and leafy shores. It is a skilled hunter that eats fish, frogs, and small animals.

© Amber Leung

Scale details

Baby cottonmouths have a bright yellow tail that looks just like a tiny worm! They wiggle it to "invite" curious animals to come closer... and when they do, the baby snake gets an easy meal.

Behavior

The Northern Cottonmouth opens its mouth wide to show a bright, white inside, just like fluffy cotton when it feels threatened.

Pop Quiz

Which snake skull is from a **venomous** snake?

A

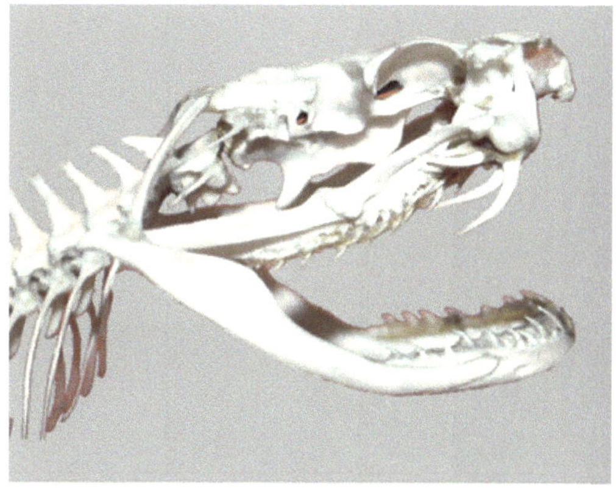

B

Answer: __________

Range: Virginia

Present

Not Present

Family	Viperidae
Average Length	30 - 48 in. (76 - 122 cm)
Record Length	74 in. (188 cm)
Reproduction	Viviparous
Other common names	Water moccasin, moccasin Water viper

Harmless

Eastern Wormsnake

Carphophis amoenus

Eastern Wormsnake is a tiny and shy snake that spends most of its life underground. They are found hidden under logs, rocks, or dirt, quietly searching for its favorite food which are earthworms. The snake is most active from spring to late summer (around March - December depending on weather).

Scale details

This little snake is smooth, shiny, and brown on top, with a bright pink belly that looks like a surprise hidden underneath. Because it's so small and slender, it's often mistaken for a worm.

Behavior

This snake is completely harmless and very gentle. Instead of biting, it tries to wiggle away or burrow back into the ground when it feels scared.

Trivia

Eastern Wormsnakes are oviparous, laying 2-6 eggs in mounds of humus or in or under decaying, moist logs. Eggs are smooth, oblong, and thin-shelled.

Wormsnake Egg
© Suzanne Collins

Range: Virginia

Present

Not Present

Family	Colubridae
Average Length	7.5 - 11 in. (19 - 28 cm)
Record Length	13.25 in. (33.7 cm)
Reproduction	Oviparous
Other common names	Blind snake/ worm, little red snake, cricket snake

Harmless

Northern Scarletsnake

Cemophora coccinea copei

Northern Scarletsnakes reproduce by laying eggs. In late summer, females lay small clutches of about 2 to 9 eggs, usually hidden in soft, moist soil or under leaf litter and decaying wood, especially in sandy pine habitats. The eggs are fairly large for the snake's size and are carefully tucked away in protected spots.

Color details

Northern Scarletsnakes is a colorful snake. When the baby snakes hatch, they already look like tiny versions of adults, though their red markings may appear more pinkish at first.

Behavior

This snake loves to burrow into the leaf litter and soil. From the moment they emerge, they are completely independent and begin life on their own.

Color time

What are the colors of the Northern Scarletsnake?
Color the snake!

R ___ ___

W ___ ___ T ___

B ___ A ___ ___

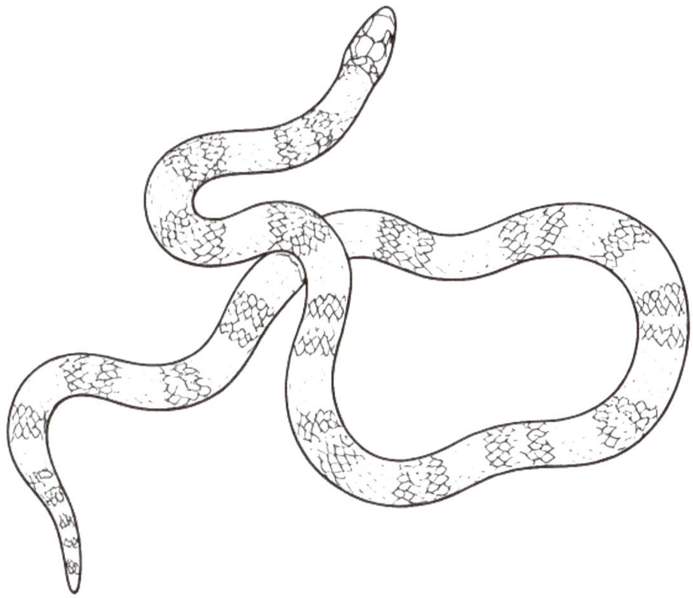

Range: Virginia

Present

Not Present

Family	Colubridae
Average Length	14 - 20 in. (36 - 51 cm)
Record Length	32.6 in. (82.8 cm)
Reproduction	Oviparous
Other common names	Milk snake, False coral snake

Harmless

Northern Black Racer

Coluber constrictor constrictor

Northern black racers are sleek and simple in appearance. As adults, their bodies are a smooth, shiny black, almost like polished glass gliding through the grass. Underneath, their bellies are a softer gray, and their chins are often white, like a small splash of light against the dark.

Juvenile

As they grow, these markings slowly fade, and their color deepens until they become the glossy black snakes we recognize. Baby racers look completely different. Instead of solid black, they have a pattern of gray or brown blotches across their bodies, helping them blend into their surroundings.

Behavior

When threatened, it quickly escapes into grass, bushes, or burrows.

Pop Quiz

Which photo shows the juvenile Northern Black Racer?

A

B

Answer: __________

Range: Virginia

Family	Colubridae
Average Length	36 - 60 in. (90 - 152 cm)
Record Length	73 in. (185.4 cm)
Reproduction	Oviparous
Other common names	American black snake, black runner, American racer snake

Harmless

Northern Ring-necked Snake

Diadophis puctatus edwardsii

The Northern Ring-necked Snake is small. It rarely grows longer than a standard 12-inch ruler. These snakes are easily recognized by their smooth, slate-gray or bluish-black bodies and a distinct, bright yellow or cream-colored "necklace" right behind their heads.

To stay safe, they never bite instead, they curl their tails to show off their bright bellies, which warns predators to stay away.s the necklace right behind their neck.

Behavior

They are shy and mostly active at night. They spend their days hiding in "secret" spots like damp leaf litter, under rotting logs, or beneath flat rocks.

Trivia

Northern Ring-necked Snake

Single nape band

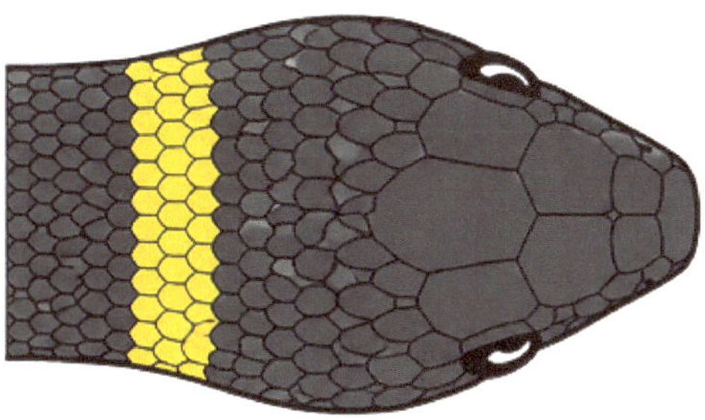

One wide yellow band across the back of the neck.

Southern Ring-necked Snake

Broken nape band

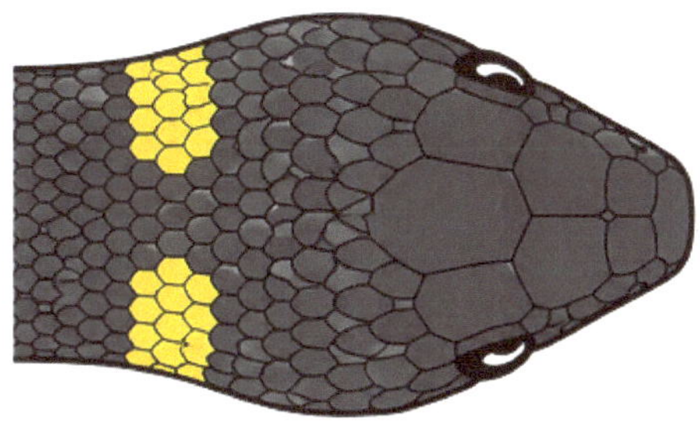

Two yellow patches with a gap between them.

Range: Virginia

Family	Colubridae
Average Length	36 - 60 in. (90 - 152 cm)
Record Length	73 in. (185.4 cm)
Reproduction	Oviparous
Other common names	American black snake, black runner, American racer snake

Common Rainbow Snake

Farancia erytrogramma erytrogramma

Harmless

The Common Rainbow Snake is one of the most beautiful and colorful snakes in North America, living a hidden life in the wetlands and sandy soils of Virginia. However, Common Rainbow Snakes are generally considered highly aquatic.

Color details

The snake have "rainbow" pattern, featuring three bright red stripes running down a shiny, bluish-black back and a yellow belly decorated with rows of black spots.

Habitat

These snakes are expert swimmers and spend much of their time in freshwater rivers, marshes, and creeks because they have a very specific favorite food: freshwater eels!

Trivia

The tail has a sharp harmless spine.

Range: Virginia

Present

Not Present

Family	Colubridae
Average Length	27 - 48 in. (68.8 - 122 cm)
Record Length	66 in. (167.6 cm)
Reproduction	Oviparous
Other common names	Horn snake, mud snake, red-lined snake

Harmless

Eastern Hog-nosed Snake

Heterodon platirhinos

The Eastern Hog-nosed Snake is the ultimate drama queen of the reptile world. When a Hog-nosed snake feels threatened, it tries to look scary by flattening its neck like a cobra and letting out a loud hiss. If act doesn't work, it tries its most famous trick: playing dead.

Color details

The Eastern Hog-nosed Snake has two common color: the dark brown or black blotches and almost completely black or very dark gray.

Food

They almost exclusively eat toads. While many animals stay away from toads because of their toxic skin, the Hog-nosed snake has a special stomach that can handle the poison safely.

Trivia

The snake will flip onto its back, open its mouth, and stick out its tongue as if it's totally gone! If you try to flip it back onto its belly, it will immediately roll back over onto its back again, insisting that it is a very convincing "dead" snake until it thinks the coast is clear.

Range: Virginia

Present

Not Present

Family	Colubridae
Average Length	20 - 33 in. (51 - 84 cm)
Record Length	45.5 in. (115.6 cm)
Reproduction	Oviparous
Other common names	Black viper/hog-nosed snake, blowing snake, Puff Adder

Harmless

Eastern Kingsnake

Lampropeltis getula

The Eastern Kingsnake is widely considered the "King of the Forest" because it is a brave hunter that can even eat venomous snakes like Copperheads! When a Kingsnake catches its dinner, it uses its strong muscles to squeeze its prey. They are immune to the venom of pit vipers allowing them to hunt other snakes safely.

Color details

They are famous for their beautiful chain-link pattern, which consists of a shiny black body with white or yellow bands that look like a chain wrapped around them.

Behavior

If they feel scared themselves, they might wiggle their tail in dry leaves to make a rattling sound or hide their head inside their own coils to stay safe.

Trivia

In Virginia, mother Kingsnakes lay about 9 to 17 eggs in early summer, and tiny hatchlings emerge already wearing their signature chain-link scales.

Range: Virginia

Present

Not Present

Family	Colubridae
Average Length	36 - 48 in. (90 - 122 cm)
Record Length	82 in. (208.3 cm)
Reproduction	Oviparous
Other common names	Black king snake, Common king snake, Common chain snake

Harmless

Northern Mole Kingsnake

Lampropeltis rhombomaculata

The Northern Mole Kingsnake is the "underground expert" of the kingsnake family. They are usually long and have a smooth, cylindrical body that helps them slide through tunnels easily.

Young Mole Kingsnake
© Graham Howe

Color details

While young snakes have bright, chestnut-brown spots on their backs, older snakes often turn a solid, chocolatey brown as their patterns fade with age.

Habitat & Behavior

Because they live most of their lives in burrows or under the soil, these snakes are very rarely seen by people. They are very gentle and harmless to humans, and they spend their nights and days digging through sandy or loamy soil in search of a snack.

Trivia

In Virginia, these snakes are most active from May to September. Mother snakes lay a clutch of about 7 to 18 eggs in the early summer, hiding them deep in the dirt where they stay moist and safe.

Range: Virginia

Family	Colubridae
Average Length	30 - 40 in. (70 - 102 cm)
Record Length	56.25 in. (143 cm)
Reproduction	Oviparous
Other common names	Blotched kingsnake, brown snake, ground snake

Harmless

Eastern Milksnake

Lampropeltis triangulum

The Eastern Milksnake is one of the most colorful and misunderstood snakes in Virginia. It got its unusual name from an old folk tale that claimed these snakes would sneak into barns to "milk" cows.

Color details

These snakes are usually famous for their beautiful checkerboard bellies and the distinct Y or V shaped mark on the back of their heads.

Food

Milksnakes are excellent helpers for farmers because their favorite food is small mice and voles. They are like the Kingsnake, they have special blood that makes them immune to the venom of Copperheads, allowing them to hunt other snakes if they need to.

Trivia

Depending on where you live in Virginia, a Milksnake might look very different! In the mountains, they usually have chocolate-brown or reddish blotches on a gray body. However, in the warmer Coastal Plain, they can have bright red, black, and yellow rings that make them look like tiny tropical treasures.

Mountain side

© John White

Coastal plain

© Jim Brighton

Range: Virginia

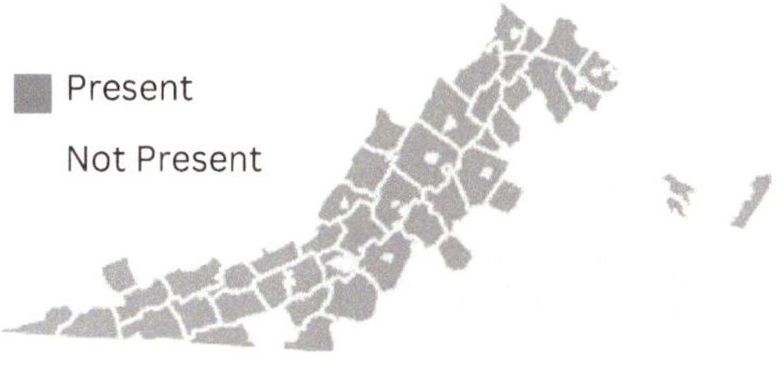

Family	Colubridae
Average Length	20 - 36 in. (61 - 90 cm)
Record Length	51 in. (132.1 cm)
Reproduction	Oviparous
Other common names	Adder, chin snake, sand-king

Harmless

Northern Watersnake

Nerodia sipedon sipedon

The Northern Watersnake is Virginia's most famous water-loving reptile. You can find them splashing around in almost any wet place, from big lakes and rivers to small backyard ponds and even roadside ditches.

Color details

They are often brown or gray with dark bands that wrap around their bodies near their heads.

Behavior

Northern Watersnakes are actually harmless to humans, but they can be a bit grumpy if picked up and will often wiggle and bite to try to get away. These snakes are amazing swimmers and hunters. They love to eat fish and amphibians, like frogs and salamanders.

Pop Quiz

Which photo shows Northern Watersnake?

A

B

Answer: __________

Range: Virginia

Family	Colubridae
Average Length	20 - 36 in. (61 - 90 cm)
Record Length	51 in. (132.1 cm)
Reproduction	Oviparous
Other common names	Adder, chin snake, sand-king

Harmless

Northern Rough Greensnake

Opheodrys aestivus

The Northern Rough Greensnake is the ultimate "hide-and-seek" champion of the forest. With its slender body and bright, leafy-green color, it looks exactly like a vine. While many snakes spend their time on the ground, this snake loves to live in trees and shrubs.

Behavior

Instead of fighting, they use their amazing camouflage to disappear into the leaves. When they sleep at night, they coil up tightly on a branch, looking just like a bunch of green stems.

Food

They are most active during the day, hunting for crunchy snacks like grasshoppers, crickets, spiders, and even caterpillars.

Trivia

Alive

© Zeev Ginsburg

Dead

© Mike Dunn

Range: Virginia

Family	Colubridae
Average Length	22 - 32 in. (56 - 81 cm)
Record Length	45.6 in. (115.9 cm)
Reproduction	Oviparous
Other common names	Brush snake, Grass snake, Vine snake, Green summer snake

Harmless

Central Ratsnake

Pantherophis alleghaniensis

Central Ratsnake is the only snake in the state that can grow to be over 6 feet long, with some record-breakers reaching over 8 feet. They are good climbers. They use their flat bellies (which are shaped like a loaf of bread) to grip bark, allowing them to scale tall trees and even brick walls in search of birds and squirrels.

© Ty Smith

Color details

© Vermont Reptile and Amphibian Atlas

Juvenile Ratsnakes are light gray with dark brown or black blotches while, the adults are mostly solid black with a white chin and a "checkerboard" belly.

Habitat & Behavior

Because they are so big and often live near barns or old houses, they are the most commonly seen snake in Virginia. They are completely harmless and very helpful to have around because they are expert at catching mice and rats.

Trivia

Juvenile

© Paul Sattler

Adult

© John White

Range: Virginia

Family	Colubridae
Average Length	42 - 72 in. (106.7 - 183 cm)
Record Length	101 in. (256.5 cm)
Reproduction	Oviparous
Other common names	Black ratsnake, Black racer, Chicken Snake

Harmless

Eastern Smooth Earthsnake

Virginia valeriae valeriae

Eastern Smooth Earthsnake are expert burrowers and spend most of their lives living underground or hiding under big logs and rocks. They are a plain grayish-brown color with shiny, smooth scales that help them slide through the dirt like a little silk ribbon.

Color details

Their tiny family usually has about 6 or 7 snakes. When they are born, the babies are only 3 inches long—about the size of a crayon—and weigh less than a paperclip.

Food & Behavior

Their favorite food is earthworms, and they spend their nights hunting for them in damp soil. They are completely harmless and very gentle. If one gets scared, it might stay very still and "play dead" or release a smelly scent to tell predators to go away.

Trivia

Adults are usually smooth gray, brown, or reddish-brown with a creamy belly.

Range: Virginia

Present

Not Present

Family	Colubridae
Average Length	7 - 10 in. (18 - 25.4 cm)
Record Length	15.5 in (39.3 cm)
Reproduction	Viviparous
Other common names	Brown worm snake, Smooth brown snake

Common Ribbonsnake

Thamnophis saurita saurita

Harmless

The Common Ribbonsnake is a super-fast, slender snake that loves the water. It gets its name because its thin body and bright stripes make it look like a long piece of ribbon.

Tail details

One of the coolest things about the Ribbonsnake is its tail. This long tail helps it balance while it climbs through bushes or swims across a pond.

Food & Behavior

Ribbonsnakes are excellent hunters, but they are very picky eaters. Instead, they love to hunt frogs, tadpoles, salamanders, and small fish. If they get scared, they are lightning-fast and will zip away into the grass or dive underwater to hide.

Trivia

Its extremely long tail inspired its scientific name saurita, meaning “lizard-like.”

Range: Virginia

Family	Colubridae
Average Length	7 - 10 in. (18 - 25.4 cm)
Record Length	15.5 in (39.3 cm)
Reproduction	Viviparous
Other common names	Brown worm snake, Smooth brown snake

Harmless

Dekay's Brownsnake

Storeria dekayi

While many snakes prefer the deep wilderness, these little guys are often found in backyard gardens, under old boards, or hiding in piles of leaves in city parks.

Color details

Dekay's Brownsnake are usually light brown or gray with two rows of small black spots running down their backs that look a bit like a tiny zipper.

Reproduction & Behavior

Dekay's Brownsnake is viviparous. When they are first born, the babies are incredibly small—only about 3 inches long, which is smaller than a crayon. These snakes like to stay warm by huddling together in old ant mounds or underground burrows.

Trivia

When they are born, they are almost solid black and have a tiny yellow "collar" around their necks. By the time they turn one year old, they lose their collar and turn their famous brown or gray color.

Range: Virginia

Family	Colubridae
Average Length	9 - 14 in. (23 - 33 cm)
Record Length	19.3 in. (49.2 cm)
Reproduction	Viviparous
Other common names	Brown grass snake, Dekay's snake, Ground snake

Glossary

Amphibian An animal that can live part of its life in water and part on land. Frogs, toads, and salamanders are amphibians.

Belly scales The wide scales on the underside of a snake's body. Snakes use these scales to help grip the ground and move forward.

Burrow A hole or tunnel in the ground where an animal can hide, rest, or hunt.

Camouflage Colors or patterns that help an animal blend in with its surroundings. Camouflage helps snakes hide from predators and sneak up on prey.

Cold-blooded A cold-blooded animal cannot make its own body heat the way people do. Snakes warm up in the sun and cool down in the shade.

Constrictor A snake that catches prey by wrapping around it and squeezing. Some harmless snakes, like kingsnakes and ratsnakes, are constrictors.

Defensive behavior Something an animal does to protect itself when it feels scared or threatened. Hissing, hiding, rattling the tail, flattening the body, or playing dead are all examples.

Egg clutch A group of eggs laid by one animal at the same time.

Fangs Special sharp teeth that some snakes use to inject venom into prey.

Habitat The place where an animal naturally lives. A snake's habitat might be a forest, field, wetland, riverbank, backyard, or underground burrow.

Harmless Not dangerous to people. Many snakes may still bite if they are scared, but harmless snakes do not have dangerous venom.

Hatchling A baby snake that has just come out of its egg.

Hibernation A deep resting period some animals enter during cold weather. Many snakes spend the winter hidden underground, under rocks, or in safe shelters.

Juvenile A young snake that is not fully grown yet. Juvenile snakes sometimes have different colors or patterns than adults.

Glossary

Nocturnal Active mostly at night. Some snakes hunt or move around after the sun goes down.

Oviparous A word that means an animal lays eggs. Some snakes are oviparous.

Predator An animal that hunts and eats other animals. Snakes are predators because they hunt animals like mice, frogs, fish, insects, and worms.

Prey An animal that is hunted and eaten by another animal. For many snakes, prey can include mice, frogs, insects, fish, salamanders, or earthworms.

Range The area where an animal can be found. A snake's range tells us where it lives in Virginia or other places.

Reptile A cold-blooded animal with scales. Snakes, lizards, turtles, and alligators are reptiles.

Scales The small, smooth plates that cover a snake's body. Scales help protect the snake and help it move.

Scientific name The special two-part name scientists use for a living thing. For example, the Eastern Copperhead's scientific name is Agkistrodon contortrix.

Slither The way a snake moves without legs. Snakes use their muscles and scales to push against the ground and glide forward.

Species A specific kind of animal or plant. The Eastern Copperhead, Northern Watersnake, and Eastern Kingsnake are all different species of snakes.

Venom A special toxin some animals use to catch prey or defend themselves. Venomous snakes use venom, but they usually want to avoid people.

Venomous Able to inject venom through a bite. Virginia has only a few venomous snakes, but kids should never touch any wild snake.

Viviparous A word that means an animal gives birth to live babies instead of laying eggs. Some snakes are viviparous.

Wetland A wet habitat such as a marsh, swamp, or muddy area near water. Some snakes live near wetlands because there is plenty of food and cover.

References

Venomous

Leung, Amber. www.inaturalist.org/guide_taxa/494732.

Seabranddesign. www.vecteezy.com/vector-art/127941-free-rattlesnake-icons-vector.

White, John. Eastern Copperhead (*Agkistrodon Contortix)*, www.virginiaherpetologicalsociety.com/reptiles/snakes/copperhead/Northern%20Copperhead%20%20(Agkistrodon%20contortrix%20mokasen)%20.jpg.

White, John. Timber Rattlesnake (*Crotalus horridus*), www.virginiaherpetologicalsociety.com/reptiles/snakes/timber-rattlesnake/sp_timberrattlesnake002.JPG.

Harmless

Adobe Images. gardenandgun.com/seven-southern-snakes-you-might-not-have-heard-of.

Baker, Charles. animalia.bio/rainbow-snake.

Brighton, Jim. https://www.marylandbiodiversity.com/record/227389

Collins, Suzzanne. www.sciencephoto.com/media/379876/view/western-blind-snake-with-eggs

Clifford, Mike. https://www.virginiaherpetologicalsociety.com/reptiles/snakes/eastern-hog-nosed-snake/Black%20hognose%20death%20feign%206.JPG

Conley, Dane. https://www.virginiaherpetologicalsociety.com/reptiles/snakes/mole-kingsnake/MoleKing_GreensvilleCounty_Conley.jpg

Dunn, Mike. https://roadsendnaturalist.com/2017/10/04/fall-herps/

FWC Fish And Wildlife Research Institute. www.southcarolinapublicradio.org/show/naturenotes/2023-11-07/young-black-racer-snake.

Gibson, Jason. www.virginiaherpetologicalsociety.com/reptiles/snakes/northern-black-racer/Black%20Racer.jpg.

Ginsburg, Zeev. https://www.inaturalist.org/taxa/29200-Opheodrys-aestivus

Howe, Graham. https://www.inaturalist.org/taxa/1304769-Lampropeltis-rhombomaculata

Klostermann, Josh. 29200-Opheodrys-aestivus

Krull, Russell. https://www.virginiaherpetologicalsociety.com/reptiles/snakes/northern-watersnake/northern_watersnake-fairfax-russell_krull.jpg

Paplanus, Peter. animalia.bio/ring-necked-snake.

Krull, Russell. https://www.virginiaherpetologicalsociety.com/reptiles/snakes/northern-watersnake/northern_watersnake-fairfax-russell_krull.jpg

Saffir, Barbara. https://www.virginiaherpetologicalsociety.com/reptiles/snakes/dekays_brownnake/storeria_dekayi-barbara_j_saffir.jpg

Sattler, Paul. https://www.virginiaherpetologicalsociety.com/reptiles/snakes/dekays_brownsnake/storeria-dekayi-with-young.jpg

Sattler, Paul. https://www.virginiaherpetologicalsociety.com/reptiles/snakes/eastern-hog-nosedsnake/Heterodon%20platyrhinos%209%20melanistic%20Campbell%20Co%202006%20copy.jpg

Sattler, Paul. https://www.virginiaherpetologicalsociety.com/reptiles/snakes/eastern-ratsnake/Pantherophis%20alleghaniensis%20eating%20Hyla%2012%20Smith%20Mt%20State%20Park%2012%20June%202009.jpg

Sattler, Paul. https://www.virginiaherpetologicalsociety.com/reptiles/snakes/eastern-smooth-earthsnake/smooth_earthsnake-ps.jpg

Smith, Ty. https://www.vtherpatlas.org/herp-species-in-vermont/pantherophis-alleghaniensis/

Smith, Ty. https://www.virginiaherpetologicalsociety.com/reptiles/snakes/eastern-smooth earthsnake/Eastern%20Smooth%20Earth%20Snake%20(Virginia%20valeriae).JPG

Vandeventer, Terry. https://www.inaturalist.org/taxa/29925-Heterodon-platirhinos

Vermont Reptile and Amphibian Atlas. https://www.vtherpatlas.org/herp-species-in-vermont/pantherophis-alleghaniensis/

Wilhelm, Mike. https://a-z-animals.com/animals/smooth-earth-snake/

Williams, John. https://www.inaturalist.org/guide_taxa/1502408

White, John. www.virginiaherpetologicalsociety.com/reptiles/snakes/eastern-wormsnake/sp_easternwormsnake001.jpg.

White, John. www.virginiaherpetologicalsociety.com/reptiles/snakes/northern-scarletsnake/sp_scarletsnake004.jpg.

White, John. www.virginiaherpetologicalsociety.com/reptiles/snakes/northern-ring-necked-snake/sp_northernringneckedsnake003.JPG.

White, John. www.virginiaherpetologicalsociety.com/reptiles/snakes/common-rainbow-snake/Common%20Rainbow%20Snake%20(Farancia%20erytrogramma%20erytrogramma)115.JPG

White, John. https://www.virginiaherpetologicalsociety.com/reptiles/snakes/common-rainbowsnake/Common%20Rainbow%20Snake%20(Farancia%20erytrogramma%20erytrogramma)095.JPG

White, John. https://www.virginiaherpetologicalsociety.com/reptiles/snakes/eastern-kingsnake/Eastern%20Kingsnake%20juvenile%20(Lampropeltis%20getula%20getula).JPG

White, John. https://www.virginiaherpetologicalsociety.com/reptiles/snakes/eastern-kingsnake/sp_easternkingsnake003.jpg

White, John. https://www.virginiaherpetologicalsociety.com/reptiles/snakes/mole-kingsnake/sp_molekingsnake002.JPG

White, John. https://www.virginiaherpetologicalsociety.com/reptiles/snakes/eastern-milksnake/sp_easternmilksnake002.JPG

White, John. https://www.virginiaherpetologicalsociety.com/reptiles/snakes/eastern-milksnake/sp_easternmilksnake001.JPG

White, John. https://www.inaturalist.org/taxa/29305-Nerodia-sipedon

White, John. https://www.virginiaherpetologicalsociety.com/reptiles/snakes/rough-green-snake/sp_roughgreensnake002.jpg

Gibson, Jason. www.virginiaherpetologicalsociety.com/reptiles/snakes/northern-black-racer/Black%20Racer.jpg.

Ginsburg, Zeev. https://www.inaturalist.org/taxa/29200-Opheodrys-aestivus

Howe, Graham. https://www.inaturalist.org/taxa/1304769-Lampropeltis-rhombomaculata

Klostermann, Josh. 29200-Opheodrys-aestivus

Krull, Russell. https://www.virginiaherpetologicalsociety.com/reptiles/snakes/northern-watersnake/northern_watersnake-fairfax-russell_krull.jpg

White, John. https://www.virginiaherpetologicalsociety.com/reptiles/snakes/common-ribbonsnake/sp_commonribbonsnake001.JPG

White, John. https://www.virginiaherpetologicalsociety.com/reptiles/snakes/common-ribbonsnake/easternribbonsnake.jpg

White, John. https://www.virginiaherpetologicalsociety.com/reptiles/snakes/dekays _brownsnake/sp_northernbrownsnake001.jpg

www.ingramcontent.com/pod-product-compliance
Lightning Source LLC
Chambersburg PA
CBHW041323210926
51847CB00002B/30
* 9 7 8 1 9 7 1 7 0 5 0 2 6 *